OUT OF THIS WORLD

Meet NASA Inventor Javid Bayandor and His Team's

Bioinspired Venus Flier

WORLD BOOK

World Book, Inc.
180 North LaSalle Street
Suite 900
Chicago, Illinois 60601
USA

For information about other World Book publications, visit our website at www.worldbook.com or call 1-800-WORLDBK (967-5325).

For information about sales to schools and libraries, call 1-800-975-3250 (United States), or 1-800-837-5365 (Canada).

© 2024 (print and e-book) by World Book, Inc. All rights reserved. No part of this publication may be reproduced, stored in a retrieval system, or transmitted in any form or by any means (electronic, mechanical, photocopying, recording, or otherwise) without written permission from World Book, Inc.

WORLD BOOK and the GLOBE DEVICE are registered trademarks or trademarks of World Book, Inc.

Produced in collaboration with the National Aeronautics and Space Administration (NASA).

Library of Congress Cataloging-in-Publication Data for this volume has been applied for.

Out of This World
ISBN: 978-0-7166-6564-9s (set, hc.)

Bioinspired Venus Flier
ISBN: 978-0-7166-6565-6 (hc.)

Also available as:
ISBN: 978-0-7166-6573-1 (e-book)
ISBN: 978-0-7166-6581-6 (soft cover)

Staff

Editorial

Vice President
Tom Evans

Senior Manager, New Content
Jeff De La Rosa

Writer
William D. Adams

Editor
Emma Flickinger

Curriculum Designer
Caroline Davidson

Proofreader
Nathalie Strassheim

Indexer
Nathaniel Lindstrom

Graphics and Design

Senior Visual
Communications Designer
Melanie Bender

Digital Asset Specialist
Rosalia Bledsoe

Acknowledgments

Cover	© CRASH Lab	24-25	© CRASH Lab; WORLD BOOK
3	© Jurik Peter, Shutterstock; © CRASH Lab	26-27	© Texas State Aquarium
4-5	NASA; © 19 STUDIO/Shutterstock	29	© CRASH Lab
6-7	© Detlev Van Ravenswaay, Science Photo Library	30-31	© Rich Carey, Shutterstock
8-9	© Droneandy/Shutterstock	32-33	© CRASH Lab
10-11	NASA/JPL	34-35	© Album/Alamy Images
13	N. Bayandor	36-37	© CRASH Lab
14-17	© Shutterstock	38-39	© Science Photo Library/Alamy Images
18-19	© CRASH Lab	40-41	© Science History Images/Alamy Images
20-21	© Nick Polanszky, Alamy Images	42-43	© Jurik Peter, Shutterstock
		44	WORLD BOOK photo by Tom Evans

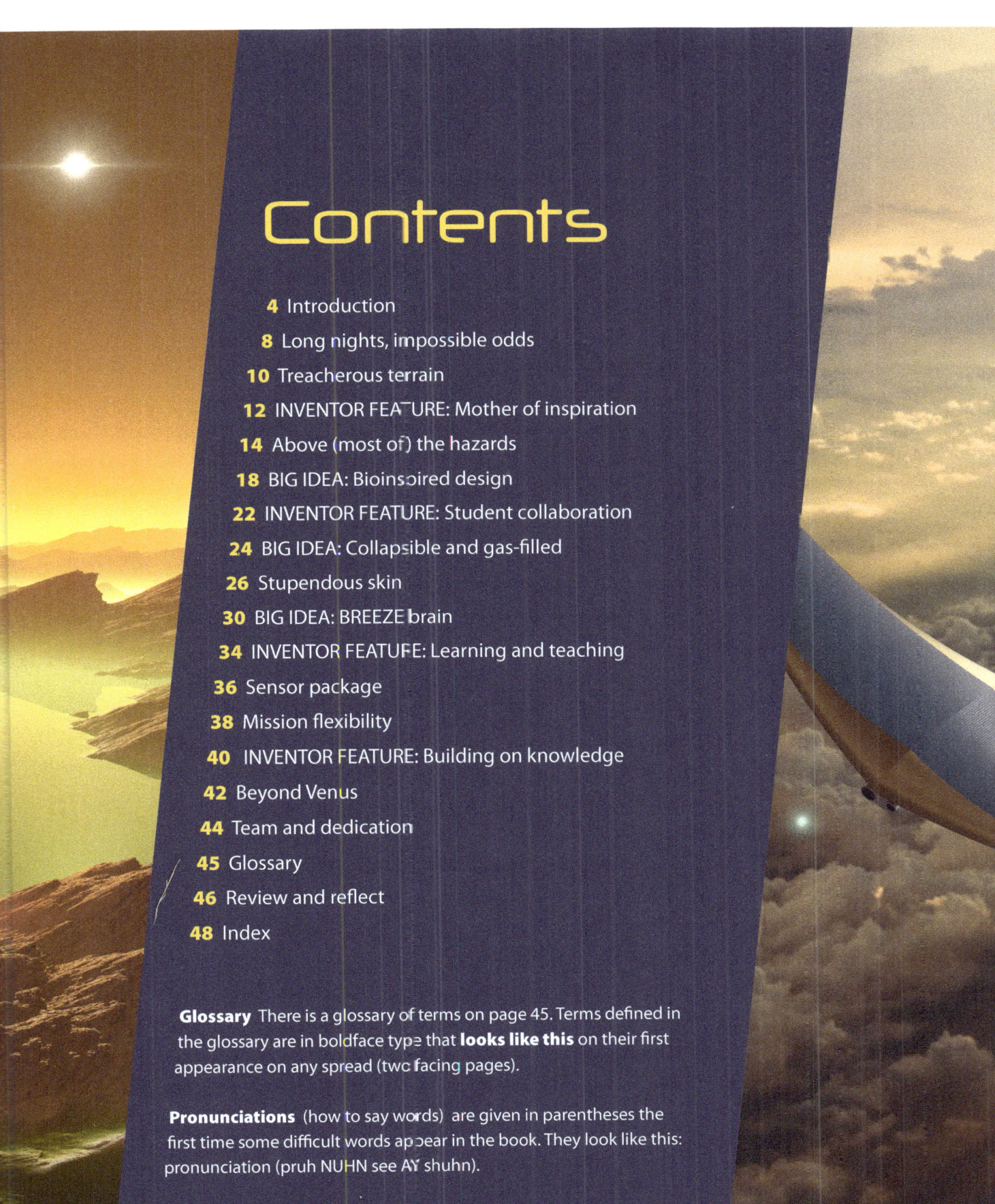

Contents

- **4** Introduction
- **8** Long nights, impossible odds
- **10** Treacherous terrain
- **12** INVENTOR FEATURE: Mother of inspiration
- **14** Above (most of) the hazards
- **18** BIG IDEA: Bioinspired design
- **22** INVENTOR FEATURE: Student collaboration
- **24** BIG IDEA: Collapsible and gas-filled
- **26** Stupendous skin
- **30** BIG IDEA: BREEZE brain
- **34** INVENTOR FEATURE: Learning and teaching
- **36** Sensor package
- **38** Mission flexibility
- **40** INVENTOR FEATURE: Building on knowledge
- **42** Beyond Venus
- **44** Team and dedication
- **45** Glossary
- **46** Review and reflect
- **48** Index

Glossary There is a glossary of terms on page 45. Terms defined in the glossary are in boldface type that **looks like this** on their first appearance on any spread (two facing pages).

Pronunciations (how to say words) are given in parentheses the first time some difficult words appear in the book. They look like this: pronunciation (pruh NUHN see AY shuhn).

Introduction

The planets Venus and Earth are similar in many ways. Venus is nearly the same size as Earth. It has a similar **orbit** around the sun. Scientists think the two planets are comparable in both makeup and general structure.

Billions of years ago, the two planets might have looked similar, too. Both had oceans of liquid water on the surface that, though very hot, could have hosted some form of life. Scientists suspect that on Venus, as well as on Earth, the surface of the planet was beginning to fracture and split, beginning the process of **plate tectonics** that cycles the chemical **element** carbon through its **atmosphere** and rocks.

But at some point—possibly as recently as 700 million years ago—Venus began to bake. **Carbon dioxide** in its atmosphere started a runaway **greenhouse effect.** The gas trapped heat from sunlight and from the planet itself like a blanket. Temperatures at the surface soared to 870 °F (465 °C). The oceans boiled away in the intense heat. Water vapor in the atmosphere *dissociated* (broke apart) and seeped into space.

Earth, on the other hand, cooled down. Plate tectonics began rearranging the surface. The action of **microbes** filled the oceans and atmosphere with oxygen, preparing the planet to support larger living things.

Could **microbial** life have developed on Venus, too, before the planet took its scorching turn?

Despite the nearness of Earth to Venus, scientists do not know all that much about the planet. Venus's dense cloud cover makes mapping the surface difficult. The extreme surface conditions make sending landing craft extremely difficult.

Engineer Javid Bayandor is working to develop a **probe** that will float above the hazards on Venus's surface. Inspired by the manta ray, the probe will glide through Venus's **atmosphere** much like that fish glides through Earth's oceans, gathering information on Earth's evil twin. His **bioinspired** Venus flier could even look for signs of life that might have taken refuge in Venus's clouds.

The NASA Innovative Advanced Concepts program. The titles in the *Out of This World* series feature projects that have won grant money from a group formed by the United States National Aeronautics and Space Administration, or NASA. The NASA Innovative Advanced Concepts program (NIAC) provides funding to teams working to develop bold new advances in space technology. You can visit NIAC's website at www.nasa.gov/niac.

Meet Javid Bayandor.

"I'm an engineer at the University at Buffalo—The State University of New York. I draw inspiration from nature to develop robotics technology. Now, I'm working to develop a flier to explore Venus."

Long nights, impossible odds

Venus presents a range of problems for potential **probes.** Many of these hazards are well known, including the high temperature and pressure at the planet's surface. But other features of the planet also make designing exploration missions difficult.

Venus has the longest day-night cycle of any planet in the **solar system.** The planet takes about 225 Earth days to travel around the sun once. But, it takes 243 days to rotate once on its axis. As a result, Venus's night lasts over 120 Earth days!

❝ In every year on Venus, you can only see the sun coming up twice—at best! ❞ —Javid

Most spacecraft get their energy from the sun—using solar power. **Landers** or **rovers** on Venus, however, would require large battery packs or alternative energy sources to make it through the long Venusian night. Furthermore, temperatures vary little day to night, so the night provides no relief from the scorching heat. A lander or rover would need to expend vast amounts of energy just to keep from overheating.

Treacherous terrain

Another lesser-known challenge to exploring Venus is the planet's treacherous **terrain.** The *tesserae* are regions where portions of Venus's inner *mantle* have been thrust to the surface. The mantle is a layer of rock between Venus's crust and core. Scientists want to deploy **landers** near the tesserae. Such **probes** would help scientists gain insight into the planet's makeup and geological history. But, because of the tectonic instability, the tesserae are full of cliffs and crags.

> ❚❚ The MAGELLAN spacecraft, launched in 1989, was able to map more than 80 percent of the surface, but the resolution is about 200 to 300 feet [60 to 90 meters]. ❚❚ —Javid

Resolution is a measure of an instrument's ability to show detail. Features smaller than this size might not appear on MAGELLAN's images. So, there are plenty of unmapped features smaller than 200 feet (60 meters) that could skewer an unsuspecting lander.

The long day-night cycle and tall surface features combine to create a third hazard: landing in the shade. Shadows shift slowly on Venus. If a lander alit in a shady spot, it could take dozens or hundreds of hours for sunlight to reach it. Such a landing could doom a solar-powered craft.

Even if a **rover** were to land in a flat, unshaded area, the extreme terrain could greatly limit its ability to roll around.

Tesserae on Venus as imaged by MAGELLAN

Inventor feature:
Mother of inspiration

Javid's mother got him interested in science and technology.

> My mother should take the credit. She spent a lot of time with me, despite the fact that she was working.
> —Javid

She taught him to read at a young age and brought home science fiction and adventure books for him to read. These works included *The Little Prince* and collections of comic strips featuring the adventurer Tintin.

The Little Prince (1943) was written and illustrated by the French aviator and author Antoine de Saint-Exupéry. In the story, a young prince from a distant planet tells the author of his experiences as he wandered among the planets seeking wisdom.

A young Javid plays with toys.

Tintin is the name of a famous European comic strip hero. Tintin was created by the Belgian cartoonist Hergé (the pen name of Georges Prosper Remi). Tintin is a reporter. The comic strip follows his many adventures throughout the world. Some of the stories have elements of mystery, science fiction, and fantasy. In one story arc begun in 1950, Tintin travels to the moon. Even though no human had yet been to space, Hergé thoroughly researched space travel and the moon to create the most scientifically accurate story possible. The science still holds up decades after the story's publication!

Above (most of) the hazards

We know that the surface of Venus is a hellish place filled with known and unknown hazards. But high above the surface, conditions become more tolerable. Just like Earth's **atmosphere,** Venus's atmosphere thins out with altitude. About 30 miles (50 kilometers) up, the pressure is the same as that on Earth at sea level. More importantly, the atmosphere's temperature at that level dips to 167 °F (75 °C). This is hotter than any place on Earth, but balmy compared to the scorching surface temperatures.

Scientists have pondered sending a balloon or airship to float above the scorching heat, crushing pressures, and treacherous **terrain** of the surface. Conditions there are certainly more comfortable, but Venus's atmosphere is not without hazards.

First, the planet's clouds are made of sulfuric acid. Sulfuric acid is highly corrosive, meaning it eats away at many materials. Any **probe** that spends time in these clouds will need a special coating to protect it.

The other hazard is the planet's extreme winds.

❚❚ There are two types of winds on the planet. One is the zonal winds, which are very fast…. They're extremely severe winds—in fact, at the higher altitudes, they're so fast that they're called *superrotational.* ❚❚ —Javid

These winds whip around the planet at about 230 miles (370 kilometers) per hour. At that speed, they *circumnavigate* (circle) Venus in about a week—60 times faster than the planet itself rotates! At first glance, it may seem that such winds could help a **probe** to circle the planet and gather data. Unfortunately, Venus's second type of winds dash such a dream.

❚❚ There are just some subtle winds longitudinally [north and south], called *meridional* winds. ❚❚ —Javid

Meridional winds push from Venus's equator to its poles. These winds are gentle, but they eventually blow anything airborne into the swirling winds at the planet's poles, called the *polar* **vortices** (singular, *vortex*).

❚❚ Any balloon or any airship you send to Venus has to spend a lot of **propulsion** to resist the 5 meter-per-second [11-mile-per-hour] meridional winds, or they are going to be sucked into the polar vortices of the planet. ❚❚ —Javid

Once a **probe** floats into one of these polar vortices, it is trapped. It could only study the environment of that vortex.

Big idea: Bioinspired design

> **"** I've always been inspired by how nature works. **"**
> —Javid

To meet the challenge of navigating in Venus's **atmosphere** and solve other problems, Javid and his team look for solutions in nature. This strategy is called **bioinspired design.**

> **"** It's quite amazing what nature does with so little. **"**
> —Javid

Of course, no animals live on Venus or in its atmosphere. So Javid and his team looked for inspiration in animals that live in comparable conditions on Earth.

> **"** We decided to look to nature and see what animal can swim or can float or can fly very seamlessly in highly fluidic conditions. **"** —Javid

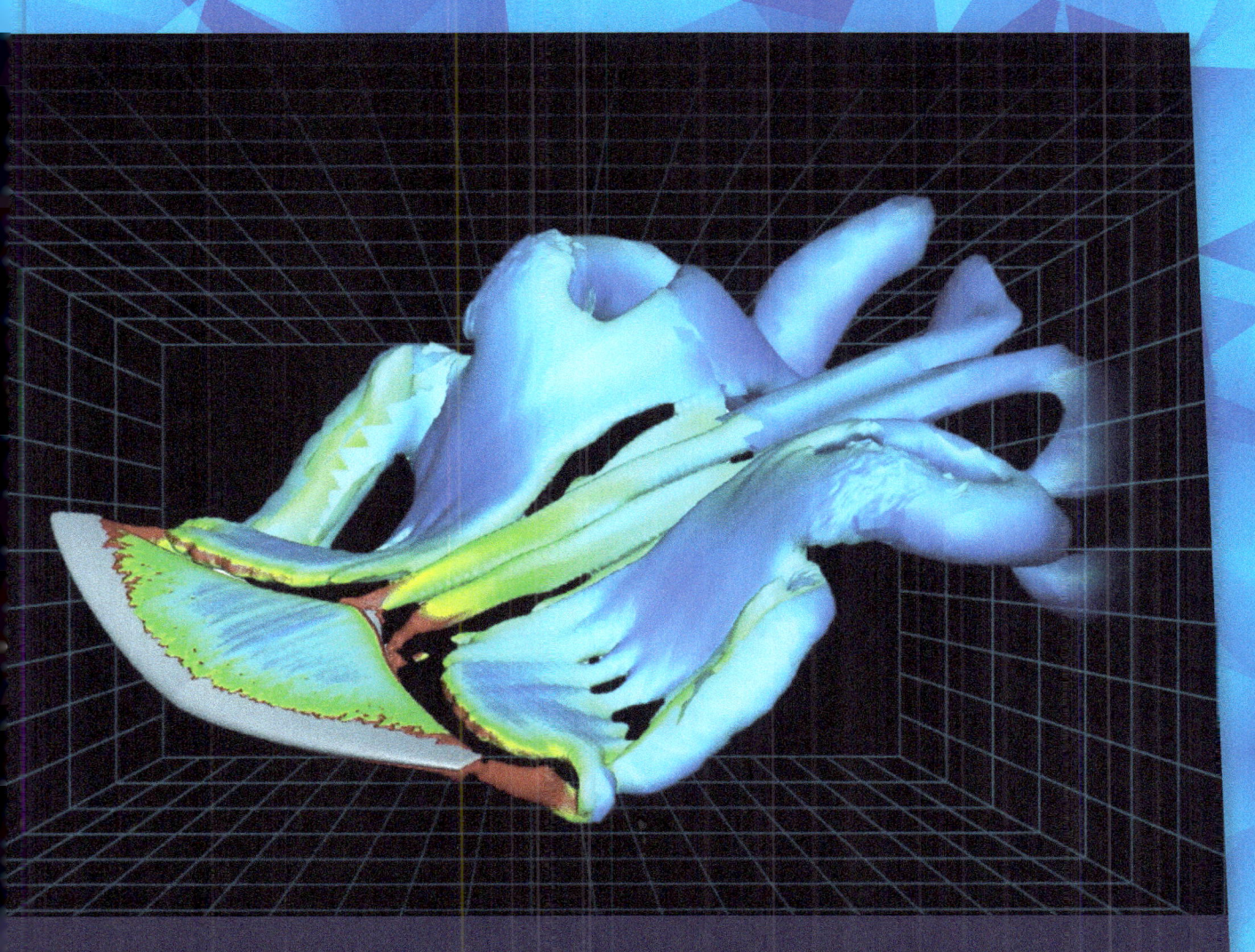

Javid and his team envisioned a craft filled with gas to make it *neutrally buoyant*. Something that is neutrally buoyant has the same density as the fluid around it and therefore remains at that level. They figured that such a craft would use less energy staying aloft. Therefore, they narrowed their search to swimming animals, which are also neutrally buoyant.

Big idea: Bioinspired design cont.

The best inspiration turned out to be the giant manta ray. The manta ray is part of a group of fish called rays. Most rays have a flat, disklike body. Among many species, the *pectoral fins* (side fins) form large "wings." Most rays swim with an *undulating* (wavelike) motion. Their movement is restricted to the edges of their pectoral fins. But manta rays swim with a flying motion, similar to the wing strokes of a bird. The entire pectoral fin is involved in the stroke. This locomotion style is extremely efficient.

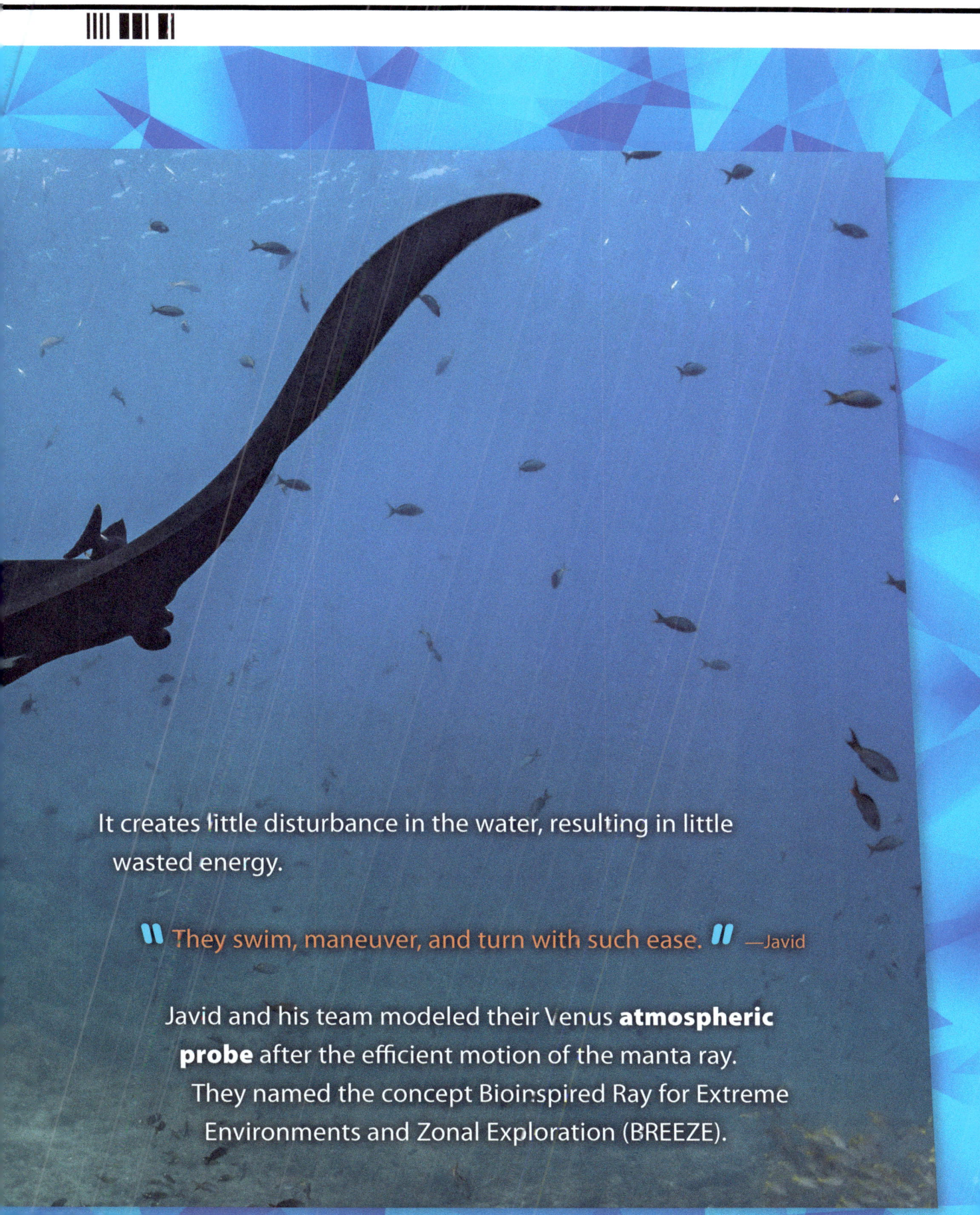

It creates little disturbance in the water, resulting in little wasted energy.

❝ They swim, maneuver, and turn with such ease. ❞ —Javid

Javid and his team modeled their Venus **atmospheric probe** after the efficient motion of the manta ray. They named the concept Bioinspired Ray for Extreme Environments and Zonal Exploration (BREEZE).

Inventor feature: Student collaboration

Javid is the founder of the Crashworthiness for **Aerospace** Structures and Hybrids Laboratory (*CRASH* Lab) at the University at Buffalo. The idea for BREEZE came about when two groups of Javid's students working on different projects came together.

One group of *CRASH* Lab students was designing a mission to the surface of Venus. Of course, they were encountering difficulty with how to overcome the high temperatures, crushing pressures, and other hazards. Another *CRASH* Lab group was designing a *soft* robot that mimicked rays, for use in monitoring coral reefs and exploring underwater environments. Soft robots have flexible structures and features, rather than the hard mechanical parts of a conventional robot. The two groups came to realize that a **bioinspired** design could work in the Venusian **atmosphere.**

CRASH LAB©

CRashworthiness for **A**erospace **S**tructures and **H**ybrids Lab

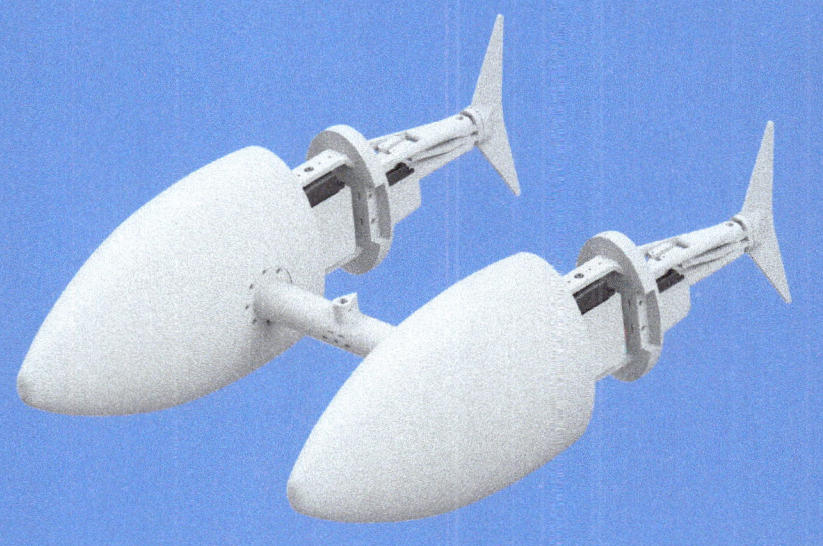

"The two teams came together really nicely. We'd been working on both projects for a few years before then."
—Javid

Rendering of a bioinspired swimming robot designed by students at CRASH Lab

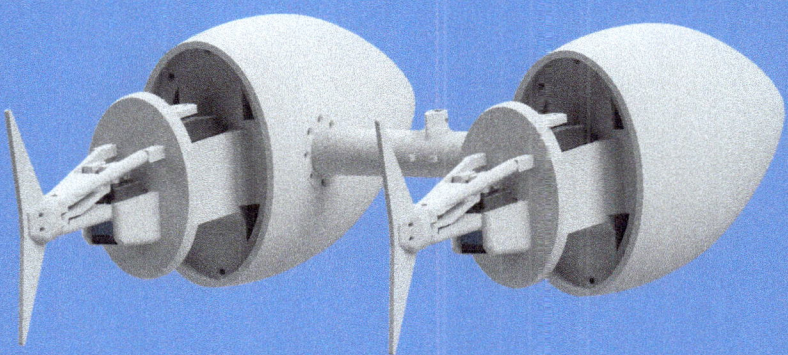

Big idea:
Collapsible and gas-filled

Like a manta ray, BREEZE will need a long wingspan. Such a shape does not lend itself well to fitting inside the **payload bay** of a rocket. Therefore, BREEZE will be stored in a folded-up configuration and deployed once it reaches the Venusian **atmosphere.**

BREEZE will inflate with gas stored in a cylinder. The gas, called *inflatant,* will both unfurl the craft and help it to float at the desired altitude.

Scientists think Venus's meridional winds vary by altitude, just like the winds in our atmosphere. BREEZE will be able to pump some of its inflatant back into the storage canister. This will enable it to adjust its buoyancy—and thus its altitude—to stay on course.

❞ When it sees that it's getting pulled off course, it can change its altitude, come down, course correct, and then reinflate itself and go the altitude that it needs to do the testing. ❞ —Javid

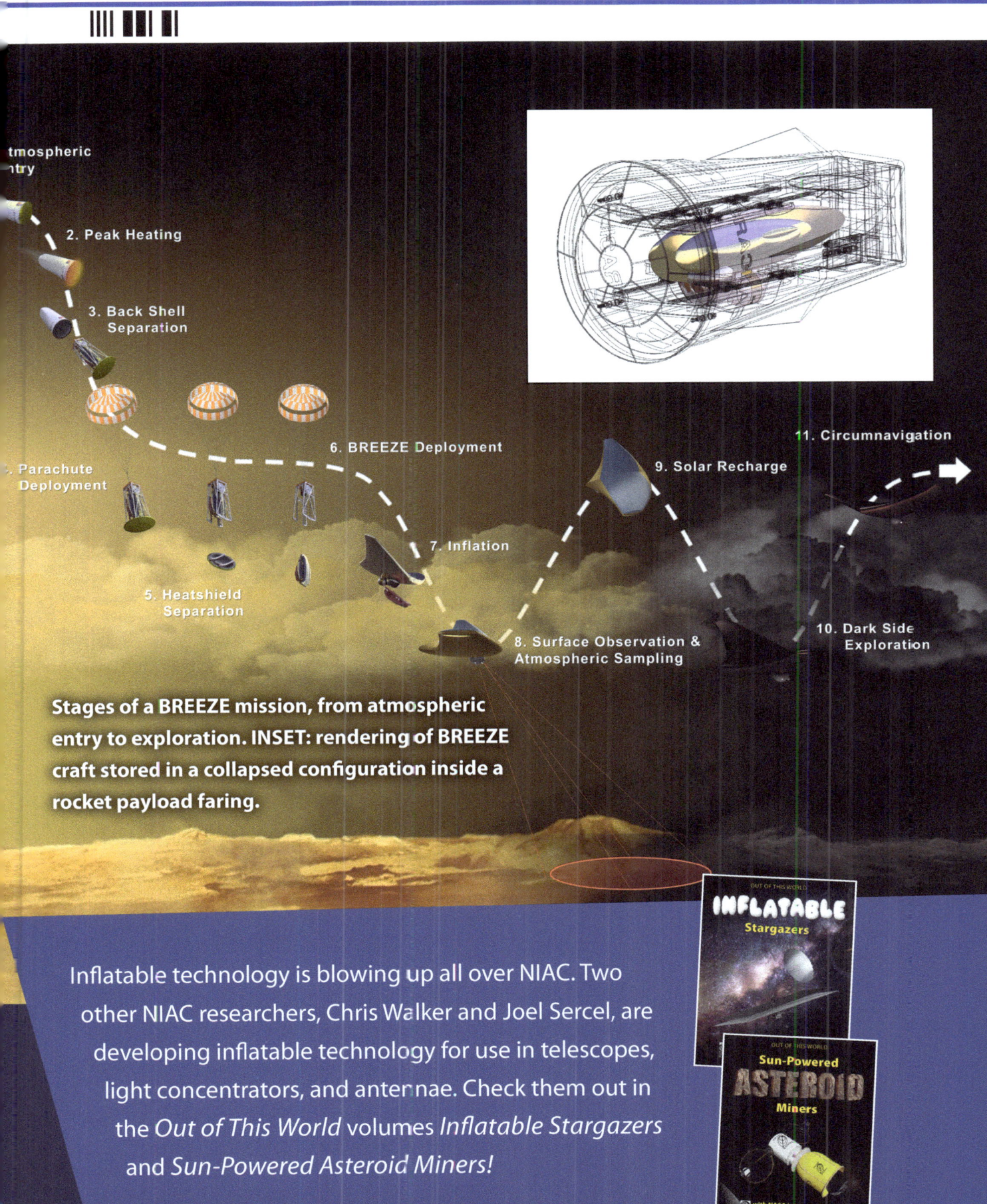

Stages of a BREEZE mission, from atmospheric entry to exploration. INSET: rendering of BREEZE craft stored in a collapsed configuration inside a rocket payload faring.

Inflatable technology is blowing up all over NIAC. Two other NIAC researchers, Chris Walker and Joel Sercel, are developing inflatable technology for use in telescopes, light concentrators, and antennae. Check them out in the *Out of This World* volumes *Inflatable Stargazers* and *Sun-Powered Asteroid Miners!*

Stupendous skin

Think about your skin. It performs a lot of different functions, such as keeping harmful germs out of your body, keeping blood and moisture inside your body, keeping you cool, and sensing the world around you. It can also heal itself, and it is flexible to allow you to move around. The skin of BREEZE will also have to excel at a variety of functions.

First, BREEZE's skin will serve as an impermeable membrane—keeping the inflatant from escaping. But that is not an easy task. Venus's sulfuric acid clouds will corrode most materials. The *hydrophobic* (water-resistant) skin of rays offers more inspiration. This skin helps these animals glide through the water more easily. Javid and his team hope to use a similar pattern within the skin of BREEZE to improve flying efficiency and reduce contact with the sulfuric acid within the clouds.

As this cownose ray lifts its fins above the surface, the water slides off them due to its hydrophobic skin.

But, **bioinspired** ray skin will not fully protect BREEZE from sulfuric acid. Therefore, the skin will contain threads of Teflon, **ceramics,** or other materials resistant to the acidic environment. The skin will have to be flexible enough to be folded up for storage and delivery and to withstand the flexing of the craft's flapping strokes.

The skin on the upper surface of BREEZE will also contain flexible solar cells. These solar cells will power the craft during the daytime and charge its batteries so it can continue to operate while it flies into darkness.

❚❚ Then I also need to use the skin's surface to be able to figure out the pressure distribution. Based on pressure distribution, I can figure out how I need to execute maneuvers to gain stability or change direction and all that. ❚❚ —Javid

The skin of animals, including humans, is like one huge **sensor** array. For flying and swimming animals, these touch sensors gather vital information on breezes or currents that help the animal stay on course.

❚❚ We need to measure all of these little pressure points on the skin to be able to figure out what is the overall **lift** or what is the overall drag on the system. ❚❚ —Javid

Simple illustration of BREEZE's proposed flapping pattern

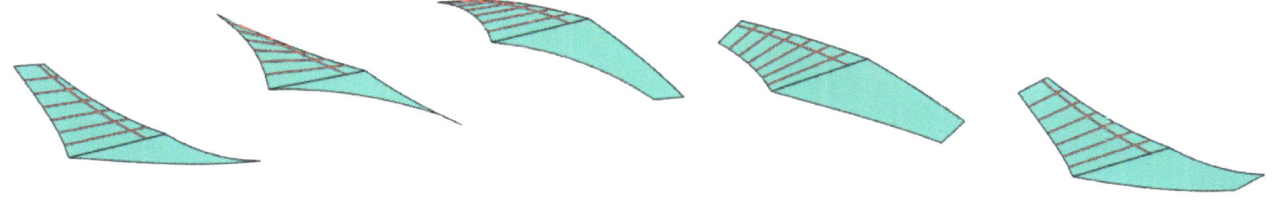

Finite element analysis showing the stressed areas of skin during different phases of the wingstroke

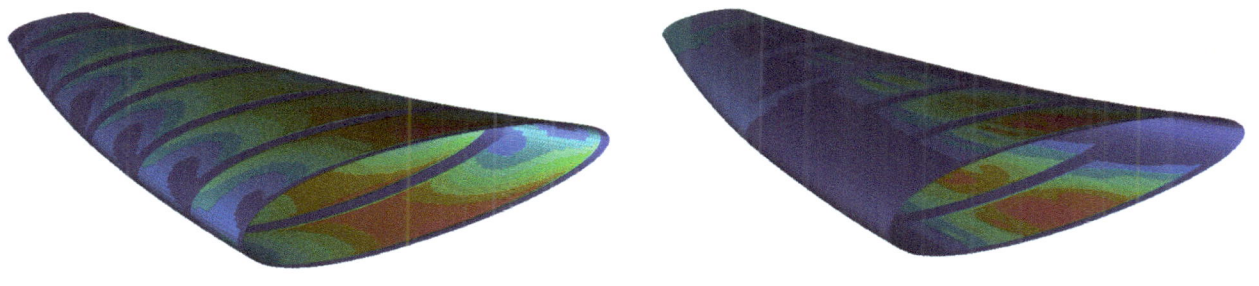

Finite element analysis showing stress across the wing ribs during flight

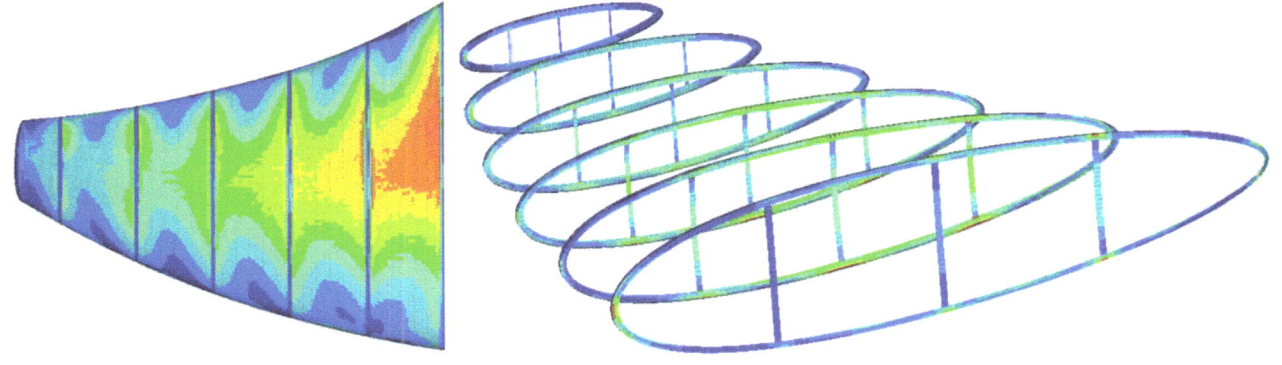

Big idea:
BREEZE brain

Developing the physical form of BREEZE is only part of the project. Imagine if mission planners tried to fly BREEZE like a remote-controlled plane. Venus and Earth can be as far as about 160 million miles (260 million kilometers) away from each other. At that distance, even transmitting at the speed of light, it would take 14 minutes for a pilot to receive data from BREEZE and 14 minutes to return commands to the craft. For flight, a 30-minute response time is way too slow!

BREEZE will thus require *artificial intelligence* (AI) to classify the situations it encounters and to act upon them. Artificial intelligence is the ability of a computer system to process information and perform tasks that would otherwise require human intelligence. There is a long way to go, however. At this stage of AI technology, Javid estimates that a robot would take more than a hundred seconds for a computer to process all the information that BREEZE would receive and act on it. That's faster than remote control from Earth, but not fast enough to keep the **probe** in a stable flight.

In contrast, a flying or swimming animal makes split-second decisions based on input from its skin, eyes, and the organs that maintain its equilibrium. Javid and his team plan to use *machine learning* to train BREEZE from many sources. Machine learning involves computer programs that learn from examples and from experience. The probe will draw on an archive of data and scenarios to understand and react to different situations on Venus.

❝ We have to estimate the next motion, because if you wait for the right answer, it would be too late. ❞ —Javid

Javid and his team will teach BREEZE to think like a manta ray.

Big idea:
BREEZE brain cont.

The team will conduct many computer simulations of the craft facing different scenarios. The results of these simulations will be uploaded into the robot's archive.

The team plans to perform wind-tunnel tests with a **prototype.** Such testing can be used to verify the computer simulations and provide additional material for the archive.

Further prototypes will be deployed in Earth's **atmosphere.** The prototypes will gather real-world data to be fed into the archive.

Another source of data will be the organisms that inspired the BREEZE design! The team will carefully record the movement of manta rays. These recordings will be used as the basis of an *algorithm* that can enable BREEZE to react

to similar situations. An algorithm is a step-by-step procedure for solving a mathematical problem in a limited number of steps.

❝ We will create a big library of scenarios and corrective strategies for it, but we also have to teach it how to go to the archive and now, using what it knows, try to remedy this scenario that has never been faced before. ❞ —Javid

Inventor feature: Learning and teaching

Javid remembers a time when he was in first grade when his teacher pivoted from a traditional lesson to engage the class.

> One time, we came to the class and she had to ask some questions or whatever and she realized that everyone was just looking down and no one was trying to get engaged. She said 'Well, I'll tell you what: I don't think this session is going to work...Let's go to the park.' And she took us to the park, and I think we found an apple tree. We sat there and she talked a little about Newton and things like that. —Javid

Sir Isaac Newton (1642-1727) was a famous English scientist, astronomer, and mathematician. He devised the theory of **gravitation.** Legend

Sir Isaac Newton

has it that Newton was sitting under an apple tree and was inspired by an apple falling from the tree to Earth.

The experience of this impromptu lesson stuck with Javid. It made him favor taking time out during teaching and learning to assess the process, make changes and adjustments if necessary, and to simply enjoy it!

Sensor package

BREEZE will carry a small **payload** module with cameras, **sensors,** and other equipment. The cameras will look down on the planet's surface to map it at a higher resolution than ever before. These maps will enable future **landers** to navigate surface hazards. They will also shed new light on the makeup of the planet.

Sensors onboard BREEZE will conduct a more detailed study of the **atmosphere.** Because conditions there are much milder than at the surface, it is possible that microbial life could exist in Venus's clouds. BREEZE could check the atmosphere for **biomarkers.** Biomarkers are signs created by the presence or activity of living things. For example, the oxygen gas in Earth's atmosphere is a biomarker. Plants and **microbes** release oxygen into the atmosphere. Few known geologic processes can produce oxygen. Furthermore, oxygen gas is highly reactive. If oxygen were not continuously replaced, oxygen concentrations would quickly fall toward zero.

Mission flexibility

The BREEZE mission will also include a parent satellite. This satellite will stay in **orbit** around Venus, taking its own measurements. It will monitor BREEZE and relay the **probe's** findings back to Earth.

The BREEZE concept—not just the craft itself—is highly flexible. Because the craft is so compact, it could be a second or third **payload** hitching a ride aboard a larger Venus mission. BREEZE could also be deployed in swarms, with numerous BREEZE craft reporting back to a single parent satellite.

Down to Earth:

Ideas from space that could serve us on our planet.

Flying is also a great way to get around Earth's **atmosphere,** too. A BREEZE-like robot could exert more control over its flight path than a traditional weather balloon. Greater control could open more opportunities for scientific and commercial missions.

Inventor feature: Building on knowledge

Javid came to appreciate all the knowledge he gathered in school as he got older. He first thought of the classes as flat, disconnected elements.

> But when you do a little bit more, instead of being a flat history, they take shape, they become three-dimensional, and you appreciate what you learned. —Javid

> So I think very important parts of your education are what you learn earlier on, but how you get to appreciate them only comes later with a bit of maturity. —Javid

Growing up, Javid drew inspiration from the American astronomer, author, and educator Carl Sagan (1934-1996). Sagan gained fame as a leading popularizer of science. He wrote several books, numerous magazine articles, and many scientific papers. He was the chief writer and

Carl Sagan filming "Cosmos"

narrator of "Cosmos" (1980), a popular public television series. The series dealt with a wide variety of scientific issues.

Javid was a young child when he became aware of Sagan, but he continues to be inspired by his work.

❝ Even now, I listen to some of his conversations and interviews and it's really amazing that he had that vision. ❞ —Javid

Beyond Venus

Aside from Venus and Earth, five other **solar system** bodies have substantial **atmospheres:** the planets Jupiter, Saturn, Uranus, and Neptune; and Saturn's moon Titan. BREEZE could be adapted to fly in any of these environments and shed more light on the structure of these bodies.

Illustration of Titan's surface

Down to Earth:

Ideas from space that could serve us on our planet.

BREEZE got its start as an Earth-bound soft robotics mission. Javid's students were designing a soft robot that mimicked rays, for use in monitoring coral reefs and exploring underwater environments. The team hopes to feed the lessons from machine learning for BREEZE back into the soft robots set to explore and protect this planet's oceans.

Team and dedication

Javid with the four CRASH Lab students who developed the initial BREEZE concept: (from left to right) Justin Page, Tyler Chau, Chet Knoer, and Alex Matta.

Other BREEZE contributors are Jamshid Samareh, Massimo Vespignani, Jonathan Bruce, Nicholas Noviasky, Pradeep Vaghela, Munjal Shah, Alexandra Nordmann, Clayton Fernandes, Trinity Blackman, Thomas Kunkel, Nicholas Deitrich, Matthew Thorton, Dylan Bellomy, Aditya Atre, Joseph Malach, Sydney Kwitowski, Sayed Asif, Edward Luthartio, Hadley Douglas, Hayley Parker, Meg Negussie, David Edwards, Matthew Montera, Robert Wilder, and Pietro Aiazzone.

❝ We always look for something that's bigger than us. Hopefully, the understanding we get from this project will help us save Earth and save humanity. We're going to have really brilliant kids who can contribute to a better future. ❞ —Javid

Javid dedicates this book to his mother, Feri:

❝ She inspired me to reach for the stars—and planets! ❞ —Javid

Glossary

aerospace the field of science, technology, and industry dealing with the flight of rockets and spacecraft through the atmosphere or the space beyond it.

atmosphere the mass of gases that surrounds a planet.

bioinspired having a design inspired by observing living things.

biomarker a sign created by the presence or activity of living things.

carbon dioxide a colorless, odorless gas present in the atmospheres of many planets, including Earth.

ceramic any of a group of nonmetal materials that are usually hard, brittle, and resistant to breakdown by chemicals and heat.

element a basic unit of matter that contains only one kind of atom.

engineer a person who uses scientific principles to design structures, such as bridges and skyscrapers, machines, and all sorts of products.

gravitation also called the force of gravity, the force of attraction that acts between all objects because of their mass. Because of gravitation, an object that is near Earth fall toward the surface of the planet. We experience this force on our bodies as our weight.

greenhouse effect the process in which certain gases (called greenhouse gases) in a planet's atmosphere trap heat that would otherwise be released into space.

lander a spacecraft designed to land on a planet, moon, or other body in space.

lift the upward reaction of an aircraft into an area of less dense air flowing over its airfoil, such as a wing or rotor blade.

microbe a very small living thing.

orbit a looping path around an object in space; the condition of circling a massive object in space under the influence of the object's gravity.

payload the useful load carried by a vehicle.

payload bay the part of the rocket set aside for carrying cargo.

plate tectonics the state of the outer crust of Earth or some other planetary body being divided into many rigid plates that collide with, sink beneath, or rise above one another.

probe a rocket, satellite, or other uncrewed spacecraft carrying scientific instruments, to record or report back information about space.

propulsion pushing something, such as a spacecraft.

prototype a functional experimental model of an invention.

rover a lander designed to move about for surface exploration.

sensor a device that detects heat, light, or some other phenomenon, producing an electric signal.

solar system the sun and everything that travels around it, including Earth and all the other planets and their moons.

terrain an area of land, usually used when referring to the land's natural surface features.

vortices (singular, vortex) swirls in a fluid.

Review and reflect

Now that you've finished reading about Javid Bayandor, use these pages to think about his experiences and BREEZE in new ways. As you work, reflect on the importance of creative problem solving, curiosity, and open-mindedness in life.

Complex problems and creative solutions

Why is NASA interested in exploring Venus?

What are some of the problems associated with exploring Venus?

How will Javid Bayandor's Bioinspired Ray for Extreme Environments and Zonal Exploration (BREEZE) overcome these challenges? What makes this solution so innovative?

Visit www.worldbook.com/resources to download sample answers, blank graphic organizers, and a rubric to evaluate writing.

Inspiration can come from anywhere!

Use a graphic organizer like the one below to map out your ideas.
What ideas or experiences led to Javid Bayandor's innovative BREEZE flyer?

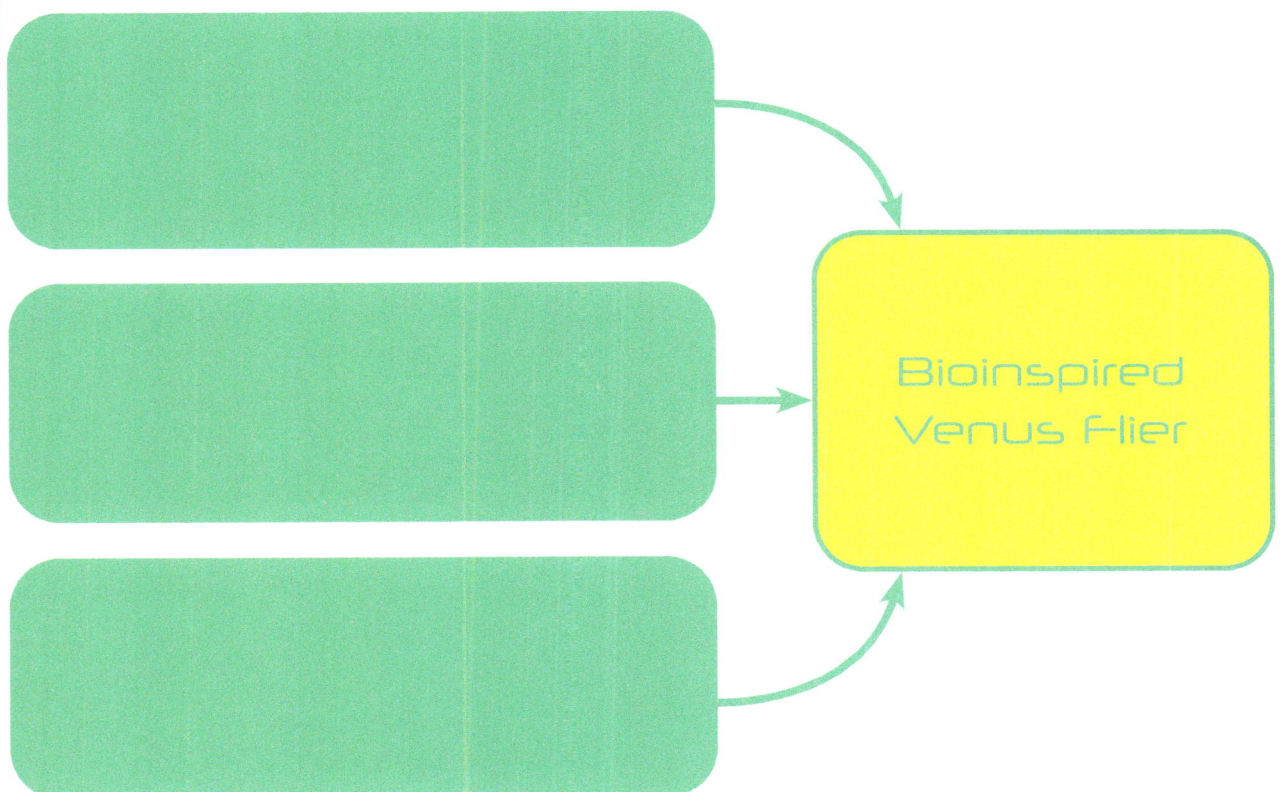

Write about it!

Think about Javid Bayandor's experiences in life and as a NIAC Fellow.

How have he and his team used creative problem solving to find success? Why might it be important to think outside the box when looking for innovative solutions?

Index

A
airships, 14, 16
algorithms, 32
altitude, 14, 16, 24
artificial intelligence (AI), 30

B
balloons, 14, 16, 36, 39
bioinspired design, 18-22, 28
biomarkers, 36
buoyancy, 19, 24

C
carbon, 4
carbon dioxide, 4
ceramics, 28
clouds, 6, 14, 26, 36
computer simulations, 32
coral reefs, 22, 43
cownose rays, 26-27
Crashworthiness for Aerospace Structures and Hybrids Laboratory (CRASH Lab), 22-23, 44

D
day-night cycles, 8, 10

F
finite element analysis, 29
fins, 20, 27

G
gases, 4, 19, 24, 36
greenhouse effect, 4

H
hydrophobic skin, 26-27

I
inflatant, 24, 26

J
Jupiter, 42

L
landers, 8, 10, 36
life, 4, 6, 36
light, speed of, 30

M
machine learning, 31, 43
MAGELLAN (spacecraft), 10-11
manta rays, 6, 20-21, 24, 31-32
mapping, 6, 10, 36
meridional winds, 16, 24
microbes, 4, 36

N
NASA Innovative Advanced Concepts program (NIAC), 7, 25
National Aeronautics and Space Administration (NASA), 7
Neptune, 42
Newton, Isaac, 34-35

O
oceans, 4, 43
orbits, 4, 38
oxygen, 4, 36

P
plants, 36
plate tectonics, 4
polar vortices (vortexes), 16
poles of Venus, 16
pressure, 8, 14, 22, 28

R
rockets, 24
rovers, 8, 10

S
Sagan, Carl, 40-41
satellites, 38
Saturn, 42
sensors, 28, 36
skin, 26-29, 31
soft robots, 22, 43
solar power, 8, 10, 28
solar system, 8, 42
sulfuric acid, 14, 26, 28
sun, 4, 8
sunlight, 4, 10

T
Teflon, 28
temperature, 4, 8, 14, 22
terrain, 10, 14
tesserae, 10-11
Titan, 42-43

U
Uranus, 42

W
water, 4, 26-27
water vapor, 4
wind, 16, 24
wind tunnels, 30

Z
zonal winds, 16

www.ingramcontent.com/pod-product-compliance
Lightning Source LLC